I0560593

I REMEMBER YOU

What we carry. What we lose.

What we remember.

Patrick J. Hughes

Copyright © 2025 Patrick J. Hughes

All rights reserved.

No part of this book may be reproduced or transmitted in any form or by any means, electronic or mechanical, including photocopying, recording, or information storage and retrieval systems, without prior written permission from the author, except in the case of brief quotations used in reviews or critical articles.

This collection of poetry reflects the personal thoughts, emotions, and experiences of the author. While some poems may contain fictionalized elements or imagined scenes, they are presented in a creative and interpretive context.

No part of this book may be used or reproduced in any manner for the purpose of training artificial intelligence technologies or systems without explicit permission.

All photographs by the author or included with permission.

Special Use Notice

The author welcomes inquiries from individuals and organizations interested in using selections from this book in support of veteran mental health, healing, or related outreach efforts. While permission is often granted for these purposes, written consent must be obtained prior to any reproduction or distribution. Please contact the author directly to discuss use. Contact information can be found in the About the Author section at the end of this book.

Your respect for the author's rights is deeply appreciated.

Hardcover ISBN: 979-8-9917021-4-0
Softcover ISBN: 979-8-9917021-5-7

DEDICATION

For those who never got the chance to say it out loud.

To the ones who've carried too much for too long,
who know the weight of silence and the cost of remembering.
This is for you.

For the ones I've loved, lost, let go, or left behind.
For the version of me who needed these words before I ever
wrote them.

I remember you.
I always will.

CONTENTS

BEFORE YOU GO ON...

I have come to understand that I am a writer throughout my life. I have also come to understand that I am not. It is an odd paradox, but I'll attempt to clarify.

I write because I must. It's how I get things out of my head. I wrote two books simply to get stories out of my mind. When I attempt to leave those thoughts, those stories, those ideas behind, they consume me. Some people give me strange looks when I tell them. But it is what does work for me, so I keep doing it.

Book writing is completely different from novel writing. To others, it will even appear completely different from who they think I am. Since I've never written about 99 percent of what is discussed here, I wouldn't blame anyone for being surprised. Even my own family didn't know I was a poet.

Believe it or not, I started writing poetry in high school. It was something that a friend blurted out in passing that set me off on it. She likely doesn't even remember saying it. I've had writing spurts ever since. Things happen. Thoughts happen. And writing happens.

Frankly, I don't know how many poems I've lost over the years. Some of it I lost intentionally. Some of it simply disappeared without my meaning to lose it. I've probably lost more than I realize. What I have compiled here are 70 poems, extracted from the many that I now have on computer.

I wrote many of these poems in a state of feeling, and many of them are poems about emotional subjects. Some are very specific. Some are interpretive and general. I did not date them or comment on when and/or where they were written, and that was intentional.

My goal isn't to lead you through a certain feeling. It's

not to explain the "right" way to read anything. I only want you to feel something. Something that you have to feel. Some of these pieces were written as I was in an inpatient psych ward. Others arrived as I rode on a train. Many were written in bed when I couldn't sleep.

I haven't greeted you or explained much about the hurt behind many of these pages. Not because I'm trying to hide anything. My name is on the cover. I just don't feel like I need to explain anything further to you. Even the About the Author at the back of the book is shorter than what I do say in either of my books.

If you'll read these pages, I believe you'll come to know a lot more about me. If only one poem in this book speaks to you, share it with someone else. Maybe they'll see something there too.

"Stories can save us. Sometimes they do."
— Tim O'Brien, The Things They Carried

I REMEMBER YOU

i remember you.

curled on the floor, drunk or dissociating—did it even matter?
you stared at the ceiling like it held answers. it didn't.

it never did.

you made deals with the dark. begged it to take you.

lied to everyone… especially yourself
that you were fine.

you weren't fine.
you called silence peace,
but it was just the absence of noise you couldn't bear.

you thought being numb meant you were strong.
it just meant you were dying slower.
you hated the mirror.

i don't blame you.

i hated it too,
but i look in it now and i see someone

who crawled out of your grave.

i remember the taste of metal.
i remember the weight in your hand.
i remember the moment you almost

made it permanent.
you didn't.

i know you thought healing was a lie.
that this version of us would never exist.
but here i am.

i don't forgive you for everything.
but i understand you.
and i'm not going back.

1

ADRIFT IN THE SEAS OF LIFE

Adrift in the ocean of my emotion,
A vessel without a rudder at the waves' whim.
Waves so high they seem to kiss the sky,
Drenching from the thrashing waves crashing over my vessel.
The vessel, battered with makeshift repairs,
Seaworthy, but just.

 It was once a Ship of the Line,
A vessel that has navigated many seas.
Adrift in the ocean of existence,
A vessel without a rudder at the waves' whim.
Waves rise and fall with their own existence,
Port to starboard, Fore to aft,
Lurking to capsize the ship at any moment,
Baring the seemingly soft hull,
A hull hardened by the sea of life.

Adrift in the sea of time,
An uncharted ship at the mercy of the waves.
Wave after wave out of control,
Scanning the receding changing panorama,
Seeing the crew sparkle and then vanish.

Time is lord of all I see,

My vessel at its mercy as well,

Dissolving as I cross the ocean,

Destined to become part of those lost by the oceans before me.

Lost in the oceans of life, emotion, and time,

A vessel without a rudder at the mercy of the oceans,

Changed forever.

THE ARCHITECT'S ABYSS

I never fit in this twisted realm,
This world bled me dry from the start,
Where the air is thick with forgotten dreams,
And shadows dance with fractured hearts.

A place where smiles are stitched with lies,
Where hope is sold on shattered streets,
Where broken minds drift like smoke,
And empty eyes avoid what they can't unsee.

Whoever's lurking in the void,
Silent as the moonless night,
Do you hear the cries of the lost?
Or do you drown them in your endless flight?

Did you weave this madness with purpose?
Did you sketch this chaos with care?
Is every tear just part of your plan,
Or do you simply watch, unaware?

I've got a question for the cruel creator,
The architect of this fractured play,
Do you see us fumbling in the dark?
Or are we just pawns you throw away?

Is this the chaos you dreamed of?
Is this the world you shaped with pride?
A labyrinth where love turns to dust,
And faith withers beneath the tide?

Each soul you crafted with trembling hands,
Now lost in this twisted masquerade,
Every breath just a fleeting spark,
Every step another mistake we've made.

I walk this path of rust and ruin,
A stranger to the place called home,
Searching for answers in the silence,
In the echoes of a world overthrown.

I never fit in this twisted realm,
This world drained me dry from the start,
But still, I wander through the void,
Seeking meaning in the darkest parts.

So tell me, creator of this endless storm,
Was it worth the blood and bone?
To craft a world of such despair,
And leave us lost, afraid, alone?

Or do you tremble in the void,
Afraid of the very thing you've wrought?
A masterpiece of endless pain,
A world where peace is never caught.

I ask you now, oh silent hand,
Is this the chaos you dreamed of?
Or do you weep for what you've done,
As we crawl through the ash and mud?

THE ROAD IN MY MIND

He'd headed west 'cause he felt that a change would do him good,

But now, even years later, he wonders what he was chasing.

California's beauty never matched the restlessness inside.

The golden hills were just a backdrop to a mind always looking for the next escape.

It's the stillness that gets to him.

The silence of routine, the comfort that starts to feel like a trap.

He once thrived in the unpredictability, the sudden orders,

The kind of movement that kept his thoughts sharp,

Where every mile meant something new, something unknown.

Now, the urge to leave lingers like a low hum in the back of his mind,

Not loud, but constant, unsettling.

It whispers of places he hasn't seen yet, roads that stretch without end,

And people—those friends scattered across the country—

Who feel like fragments of a past life,

Tied together by the moments shared in motion.

Sometimes he imagines the relief of simply leaving,

Driving until the doubts fall behind,

Reconnecting with those who understand the need to keep moving.

Each visit a reminder that he's not alone in this unsettled state,

That others, too, are scattered like him,

Drawn together not by place but by the shared experience of never quite belonging anywhere.

But the thought of staying—of roots, of routine—

It digs at him, too, making him question if it's fear of the unknown that holds him back,

Or something deeper, something he's not ready to face.

For now, it's easier to stay in place,

Though the road always seems just a thought away.

WALKING WITH DEMONS

I think it's time to walk with my demons
before I walk with the gods.
Beneath the vast, unyielding sky,
the horizon where the sea kisses the endless blue,
a place where the world narrows to a line.

I stand on the edge, the deck beneath my feet,
where the salt air stings my face
and the wind howls through the steel and cables,
carrying whispers of forgotten fears and silent grief.
Here, shadows are more than mere tricks of light.
They are memories, specters from the past,
lingering in the corners of the mind,
echoes of engines roaring and lives intertwined.

To walk with my demons is to remember,
to feel the weight of duty, the pull of the past,
to face the waves that crash within,
the turmoil of loss and the storm of regret.
It is to stand vigilant in the dead of night,
eyes straining against the darkness,
listening to the deep, resonant hum of the ship,
a lullaby of steel and ocean.

In the solitude of the watch,
there is no escape from the inner tempest,
no place to hide from the phantoms
that rise with the moon,
drawn out by the quiet
and the solemn procession of stars.
These demons are comrades in a silent march,
their presence a testament
to the unseen battles fought,

the wounds that bleed without a sound.

Before I walk with the gods,
I must navigate this shadowed path,
honor the ghosts that share this journey,
embrace the scars that tell my story.
For in the depths of these waters,
where the abyss meets the light,
I find the strength to rise,
the resilience forged in fire and salt,
a heart tempered by the vastness of the sea.

Only then, when I have walked
this haunted trail,
can I hope to stand before the gods,
with the weight of the world
eased from my shoulders,
ready to face the dawn
with the wisdom of the night.

For it is in the company of demons
that we learn to be human,
to understand the fragility and the strength,
to find peace in the turmoil,
and to embrace the silent victories
that light the path ahead.

WHISPERS AND SHADOWS

I don't listen to the voices in my head.

How can I know if I'm feeling like myself, If I'm unsure who I am? In a world that rushes by, I feel stagnant, watching the seasons flee. I want to live, not just exist.

Can I see the light in the shadow's grip? Can I feel the fight in every stumble and slip?

Through the darkness, I'll find my light; with every fall, I'll gain the strength to fight. I carve my path, no matter the pain in sight. I bend, I break, I find my strength, I sink, I soar, I rise once more. I face my fears, fight my battles, The whisper screams, "Don't follow the pack." In the silence, I'll find the strength to attack.

My mind's a maze of darker days, I wander, lost and weary, through shadows' fold, my story's told in whispers, cold and dreary. A ghost inside, where fears reside, My smile's just a façade.

Seeking peace, a sweet release, I'm flawed yet still I nod. Can I break the chains that hold me down? Can I face the pain without a sound?

I drink myself numb so I don't feel a thing, tethered to a carousel of pain I can't break free. I've lost count of the times I've nearly dimmed my light, wondering if I'll make it through the night. I write a note to tell them why I went to meet the angels. It's a chronicle of struggle, my life's entangled angles.

Just beneath the surface, my war wages, Unseen battles every night. The scars run deep, they creep and seep, out of my mind but never out of sight. The reflection lies, it disguises the despair. The dark leaves its mark, it carves it deep, I am alone.

In the quiet, I stand tall, the whispers fade away. No longer bound, I've found my ground, in darkness I won't stay. The battles fought, the lessons taught, have shaped who I've become. From shadows cast, I rise at last, my strength has just begun. In the end, I've found my strength, I'm no longer ruled by fear.

WRITING IS MY THERAPY

I need to get these jumbled thoughts
Out of my head.
My musings, my daydreams, my journals.
Pouring my heart and soul out
With every keystroke.

I need to clear my mind.
I need to regain my focus.
I... I...
Wait—what was I needing?
Oh yes...

I need to do laundry.

But first, perhaps a cup of soda.
Wait, did I pay the electric bill?
Maybe I should check my emails.
Or finally fix that squeaky door hinge.

Distractions, distractions everywhere.
My mind's a circus, and I'm the ringmaster
Who's lost control of the show.
I sit to write the next great novel,
But end up Googling "how to fold a fitted sheet."

Sigh.

Maybe enlightenment can wait.
After all, those socks won't pair themselves.
And who knows, perhaps amidst the detergent suds
I'll find the clarity I seek.

Until then, I'll jot down these thoughts—
Well, right after I switch the laundry

EMBERS OF WHAT WAS

There's a quiet storm that lingers now,
a strength that once simmered beneath the surface,
contained but powerful,
like a fire that knew its own limits
but dared to burn brighter.
Love once spoke loudly,
showing itself in the smallest ways,
in the brush of hands,
in the spaces left behind
when voices wavered.

Laughter once lit up rooms,
turning everything upside down,
worn like a crown,
helpless in its glow.
But beneath it, there was depth,
something wild and untamable,
a heart as vast as the ocean,
yet calm when needed most.

Sometimes, the shadows linger,
leaving questions of how such gentleness
could carry such fire,
how laughter cracked open the dark,
and smiles—
smiles that once were the sun,
pulling everything back, even in moments of drift.

A soul, forever at sea,
once anchored here,
finding a place amidst the endless horizon.

No hero was claimed,
but somehow, protection was offered,
not with walls or shields,
but with the quiet assurance
that even in weakness,
there was strength,
and that was enough.

Wildness lived in passion,
flaring untamed and gorgeous,
but it never scorched—
it only warmed, only filled.
Once fire,
and sparks that danced alongside,
caught in the heat,
but now, only embers remain.

ECHOES OF GOODBYES

We said our goodbyes,
With a joke and a smile,
Never thought it'd be the last,
We'd see each other for awhile.

I see you in my dreams,
Hear your voice in my head,
But when I wake up,
You're not here, it's just emptiness instead

I wish I could turn back time,
And change that fateful flight,
But all I'm left with,
Is this pain and this fight

I try to move on,
But the memories linger,
I see that plane go down,
In an endless mental flicker.

EMPTY

Emptiness sits beside me, a shadow on the passenger seat, as I drive my boys to safety, through days that stretch too long, and nights that come too soon.

She still wears the mask, still plays the role, pulling at strings that unravel me, until there's nothing left to tie, nothing left to fight, and I wonder if she even sees the damage, or if she's too lost in her own story, a story where she's always the hero, and I am just a villain in her way.

Her words spin, twist, and bite, the echo of a voice that used to be home, now a hurricane I can't escape, a storm that I brace for, again and again, no shelter, no calm, just endless waves of blame.

But I hold onto this wheel, knuckles white, eyes forward, with two little lights in the backseat, that keep me moving, even when I'm tapped out, even when the road seems endless, and my mind screams for rest.

I watch them through the mirror, their laughter like a spark in the dark, a reminder that the fight is not for her, but for them, for the smiles that cut through the fog, for the moments when they look at me, and I see trust, I see love, I see the reason I keep going.

I'm a mountain worn down to a hill, yet here I stand, empty, drained, but still here. Still driving them through, still their steady hand, even when mine trembles in the dark, even when the weight of it all feels too much to bear.

I am tired, worn thin, but I am their anchor, their shelter, and in the quiet moments, when they lean on me, when they rest their heads, I find just enough strength, to face another day, to keep the shadows at bay, to keep holding on, for them, and maybe, one day, for myself too.

PHOTOS ON THE WALL

Photos on the wall,
Echoes of a life I can't recall.
Fragments of moments, both bitter and sweet,
In a tapestry woven with memories I'd rather delete.

Photos on the wall,
Of times I'd rather forget.
I can't pick and choose
Which memories to keep, which to lose.

Should I cherish the good,
Or dwell on the bad?
I'll keep the photos on the wall,
But wish to erase the ones in my head.

Mentally fractured,
My mask worn thin,
Still struggling, still surviving,
As I fall apart within.

I fake the smile,
Trapped, going nowhere,
In a life that's trying to break me,
With burdens too heavy to bear.

But even in the darkness,
A flicker of hope remains,
A stubborn light that refuses to dim,
A whisper that calls me by name.

So I'll leave the photos on the wall,
A testament to the pain and the grace,
And while I may falter, stumble, and fall,
I'll rise again, finding strength in this space.

WHAT THE SILENCE DOESN'T SAY

Still quietness under night's stars
Quiet reflections beating on the shore.
Smoke and smoldering orange rise
into darkness—of course they do.

The fire rages in pops and breaths,
telling all we didn't.
Words we'd hidden deep in sound
rise up like they're looking for second chances.

The wind behaves as if it doesn't hear.
It's practiced at that.
So are we.

But maybe that's the point—
to stay here long enough
that silence no longer is a failure
and starts to sound like peace.

Maybe not.
But it's warm, and nobody's screaming,
and the stars don't ask questions.
So for the moment, that's enough.

HIDDEN

No one knows how disappointed I've been, How many times I've stayed in my room, staring into nothing, Allowing the tears to fall, unseen, unheard.

No one knows the times I lost hope, The times I've felt the delicate edges of myself break, When darkness closed in and I struggled to keep it away.

No one knows how many times I've had to hold it in — Swallowed the screams, concealed the weight of it all, Pretended to be whole while falling in pieces.

How many times bad thoughts have spoken their poison, And how I've fought them off, quietly, alone. Every breath a fight, every day a war, waged without honor or medals.

Nobody knows how many nights I lay awake, With my thoughts running in circles, reliving all the mistakes, All the things unsaid, all the seconds lost between my hands.

There are mornings when getting out of bed feels like a feat, When breathing is victory over blackness. Mornings when I wear a mask of smiles, deceiving the world that I am okay, While on the inside, I'm piecing myself together, bit by bit.

Sometimes the boldest among us are those who love in spite of their scars,

Who wipe away our tears in the dark so the rest of us don't have to,

Who carry the loads that would crush everyone else.

I remain whole for the people I care about, even if the parts do not fit, I wear my scars as shields, not to expose but to survive. I fight for light, even when darkness consumed me, I choose love, even when the world comes to an end.

Only the survivors of the night can understand the beauty of dawn— The small flame that slices through, so gentle and hesitant, yet sufficient to convince. The dawn reminds you that there is hope and a reason to live, to continue on, Even through the darkest nights, the sun will break, and so will you.

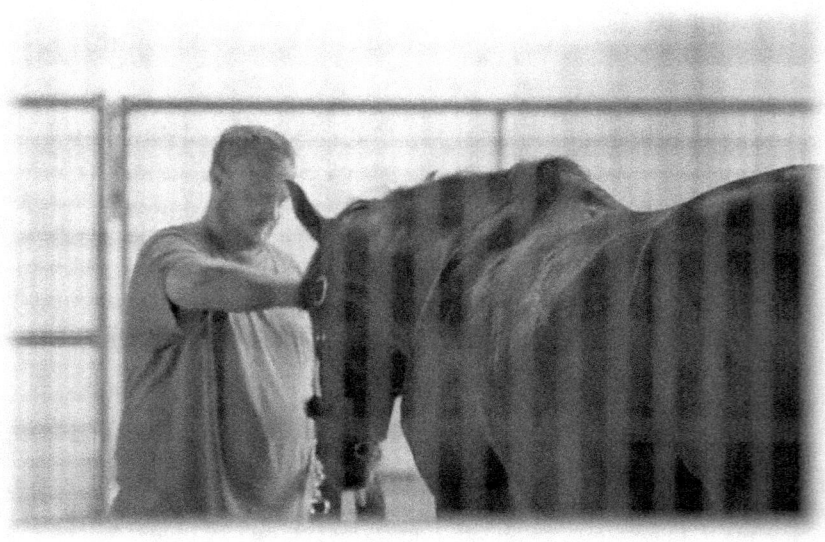

A WAR WAGES

How do you know if you're feeling like yourself
if you don't know who you are?
I've lost track of how many times
I've almost extinguished my own light.
Will I come out okay on the other side?
My memory is so lousy,
I'd forget my name if it wasn't tattooed on my arm.

I tried, I failed, I tried again.
I live in the silence,
my thoughts are deafening.
I just start a fight so I feel something.

My mind is a maze of darker days.
I wander, lost and weary.
Through shadows' fold,
my story is told in whispers,
cold and dreary.
A ghost inside, where fears reside,
my smile is just a façade.
Seeking peace, a sweet release,
I'm flawed, yet still I nod.

Drink myself numb so I don't feel a thing.
It is a carousel of pain,
I'm tethered, can't break free.
In a world that rushes by,
I'm stagnant, watching the seasons flee.
Write a note to tell them why
I went to meet the angels,
a chronicle of the struggle,
life's entangled angles.

Beneath the surface, a war wages,
unseen battles every night.

21

Scars run deep. They creep and seep,
out of my mind but never out of sight.
The reflection lies.
It disguises the despair that is set in stone.
In the dark, I leave my mark.
I carve it deep, alone.

How do you know if you're truly alive
when you're just surviving day to day?
I am in a search for warmth
in a world so cold.
I guess I'll find my own way.

ECHOES IN THE QUIET

In the quiet spaces between breaths,
You held the world with quiet might.
A soul unbending in the face of storm,
Like a mountain against the night.

Through whispered winds of fleeting time,
Your courage grew, though all seemed still.
In every step, you bore the weight,
With grace, with strength, with iron will.

The battle waged within your bones,
Yet never once did you betray
The depth of heart, the light within —
A warrior through every day.

Though shadows tried to steal your flame,
You stood like stone, unshaken, tall.
In stillness, there was endless power —
The truest strength within us all.

Now the stars will know your name,
Etched in the sky where brave hearts soar.
And we will carry you with us,
In our hearts forevermore.

A GAME OF ECHOES

Rules are written in ink,
but one hand holds the eraser.
A contract is read,
then rewritten in silence.

A game with no scoreboard,
no referee,
no end.
Only turns.

Words land clear,
then return distorted.
Meanings untangle,
only to be rewoven.

Moves shift without warning,
victory declared before the match begins.
Reason reaches forward,
but the definition changes mid-sentence.

Silence is not empty.
It is full of unspoken words,
accusations left implied,
conversations never started,
so they never have to end.

A question remains:
Does the game continue
if no one agrees to play?

WHAT REMAINS

One minute you're here, next minute you're gone.
No warning, no reason—just absence.
A gap where you once were, still occupied like you,
but empty in ways I can't define.
Memories don't ask for permission.
They pop up in the silence,
in a song, in an odor,
in the way a stranger stands with hands in pockets.
Some days, they drop like a whisper.
Others, like a blow.
Neither last very long,
but both leave a mark.
I don't revisit the past in order to simplify.
I don't assume that it was completely bad or completely good.
It was what it was, and it is,
keen at some points, tender at others.
But I don't bear it the way I used to.
Not as a scar, not as a weight.
It's just there, part of the air I inhale.
And I forgive it all.

DISTANT BUT HERE

I know I've been distant,
but I want you to know I'm still here.
It's not that I don't care,
I just haven't been reaching out.

I've been fighting my own battle,
one that feels personal,
and sometimes it's easier
to face it on my own.

I've pulled away,
isolated myself,
but I never intended
to make you feel forgotten.

I'm still here,
and when I'm better,
I'll come back.
For now, I'm trapped
in my own struggle.

It's not my intention to hurt you.
I won't be gone forever—
I just need time
to feel like I belong again.

STORM PREP

They sent us west to wait for weather
that never showed up.
Packed the jets like we were evacuating Earth,
then parked them in a borrowed hangar
to collect dust in style.

No real plan. No real danger.
Just orders from someone
who read a forecast
and wanted to be seen "doing something."

We sat for days,
guarding machines that didn't need guarding,
watching rain soak the tarmac
in the place we ran to escape it.

No tools.
No missions.
No reason.

Just long hours, bad coffee,
and an impressive amount of grown men
figuring out new ways to be bored.

Someone's pet died back home
that was the emotional high point.
Not mine, thankfully.
But the grief was real.
Mostly because it broke up the week.

Eventually we got sent back.
The storm forgot us.

Or never cared.
Like most things.

But hey
for a few days,
we got to pretend we mattered.

In Ohio, of all places

RAMP TIME

They said it was just a cargo ramp, a patch of metal bolted onto a bird that hauled mail, spare parts, and sweaty Sailors.

But out there, with the sky pouring into the cabin like something alive, it was something else.

Feet swung loose into the slipstream, boots tracing lines through a thousand feet of nothing. No speech. No headset chatter. Just the hum of turbines behind me and a blue that stretched so wide it felt like it might swallow the noise too.

The wind grabbed at my shirt, but not like it wanted to steal anything, just remind me I was still tethered to something.

This wasn't a view built for postcards or polished words. It was a view made for silence, for knowing you'd never be able to explain it to someone who hadn't sat there, knees at the edge, not thinking of war or home, just the shape of the world from that spot.

A piece of metal in the sky, and for a minute, it felt like church.

TWIN CARRIERS

Steel city on the water, twin to mine,
You drift beside us, braced against the weight
Of what comes next. The sky is clean, for now,
But smoke will follow. We all know the drill.

Your deck is bare, your birds still sleeping tight,
But not for long. The launch crews stretch and wait.
Some kid, eighteen, will pull the handle soon
And loose a fire he won't ever forget.

We're not alone. That's what this photo means.
Two ships. One war. Same orders in our hands.
Below the calm, the engines churn and groan,
A low and constant song of something grim.

I watch your flight deck crews in mirrored steps
Same helmets, jerseys, harness, colored vests.
The rituals repeat with practiced hands,
As if we train to keep the world from split.

This ocean isn't wide enough for doubt.
We know our place and purpose by the noise.
A catapult could split the morning sky
And no one here would flinch or turn away.

The water's flat, the light is gold and still.

It almost dares us to believe in peace,

But war's already echoing below

In chains and chalk lines, wings and waiting boots.

There's comfort in formation, edge to edge.

Your ship, your crew, your rhythm match our own.

We speak no words across the salted space,

Yet everything we need to say is clear.

DET LIFE

Another det.
NAS Jax this time,
again.
Same hangar, different week.
Same tools,
same gripes,
same jet with a fresh new problem
just waiting for me like it's personal.

Unpack the gear,
sign for the van,
pretend the barracks don't smell like mildew and broken
promises.

Fix it, fly it,
write it up,
rinse, repeat.

Then it's four days on the boat—
whichever carrier drew the short straw.
Different ocean, same nonsense.
No rack curtains, mystery meat at midrats,
and someone always finds a way to lose a tool in a place that
defies logic or physics.

It's Groundhog Day
with more hydraulic fluid
and slightly better coffee
if you hit the galley at the right time.

They call it training,
but I stopped learning anything new
three detachments ago.

Just keep it running,
keep it safe,
keep from stabbing someone in the QA office
before we go back to Norfolk
and pretend it wasn't exactly
like the last one.

STEEL WAKE

Steel walls echo with quiet steps,
Each hatch a mouth I've fed before.
The lights hum secrets no one kept,
I march a path, both job and chore.

Each hatch a mouth I've fed before,
Faint oil and sweat still stain the air.
I march a path, both job and chore,
Same boots, same bulkhead, same blank stare.

Faint oil and sweat still stain the air,
A low thud hums beneath my feet.
Same boots, same bulkhead, same blank stare,
Another shift, another repeat.

A low thud hums beneath my feet,
The ship breathes back, half in a dream.
Another shift, another repeat
The sea is close, but never seen.

STRAITS

Steel deck cold beneath my boots
The ship pulls a quiet halt
Mountains cut the sky in sharp shapes
White streaks sit on their ridges

Twenty-two years ride my shoulders
Shifts stacked where sleep should lie
I stay upright instead of stretching out
Eyes grainy, thoughts loose

Salt air threads the fur on my hood
Sunlight spills a silver path across the water
Could I walk that stripe of light?
What piece of me would make the far bank?

You might see calm, a postcard hush
I feel engines cooling behind steel walls
Ten borrowed minutes, no orders, no noise
This view costs sweat and hours no tourist pays

Questions surface:
How wide is a life?
Does a wake matter once it folds?
Answers sink with the ripples

Soon radios will crack, jets will scream
I slide the scene into an inner pocket
Turn back to the grind, body heavy
Mind carrying fresh weight

If you picture snow or smoke
Hold your image against the dark
Let the reflection speak for you
A pause can hold an ocean

SOUTHERN PAUSE

Cold alloy steadies my stance
Water stretches like pressed foil
Snow-capped ridges shadow the sky
A lone stripe of light skims the surface

Eyes burn from thin sleep
Hands pocket unpaid hours
Silence settles where noise belongs
Breath lifts, breaks, fades

The narrow pass feels close, then vast
Distance hides between slopes
Time drifts beyond the rail
Too wide to measure, too quick to keep

What trace survives when ripples fold?
Which questions linger once they sink?
I watch them slide below the sheen
Leaving faint rings, then nothing

Tasks wait behind the bulkhead of night
I turn, weight returning to my frame
The view stays folded in a secret sleeve
Steady as the cold, patient as the tide

TIGER GUIDE WITH NO TIGER

Officially a Tiger Guide,
with no Tiger.
Scheduled to leave early,
but plans change
like everything else in this floating mess.

I'll smile, show the guests around,
talk about how the ship runs,
how the crew functions,
what life is like out here.

But I won't say
that I sleepwalk through days,
count hours backwards,
or that my motivation
escaped weeks ago
and no one filed a report.

I'll skip over the politics,
the pettiness,
the way they hand out blame
like candy
and hoard praise like ammo.

I'll be polite.
I'll say "this is where we work,
and that's where we eat,"
and maybe leave out
that the real work
happens in silence, in sweat,
in quiet mutiny
behind locked shop doors.

THE BOLT

They spent eighteen hours
looking for a bolt.
I found it in ten minutes.
They spent four hours
on a job I closed in one.
It's not magic.
Not luck.
It's paying attention
while everyone else is posturing.
But tell me again
how I don't know what I'm doing.
Tell me again I'm a burden
while I hold the place together
quietly,
efficiently,
invisibly.
Go ahead.
I don't need recognition.
Just space.
And maybe one honest "thanks"
before I stop giving a damn.

SICK CALL SUNDAY

No sick call on Sunday.
That's the policy.
So I spent the day sweating out poison
on a steel deck hot enough to fry
whatever's left of my patience.

Corpsman said I was faking,
until I sprayed the passageway.
A three-bag IV apology followed.
She was hot, for the record.
Didn't help the fever though.

Knocked out, half-conscious,
back again the next day.
Threw up on another corpsman.
Got to see the doc this time.
They named my disease after a rat.
Makes sense. This place feels like a cage.

By day three I was curled
on the shop floor, 101 temp,
no SIQ, no mercy.
The boat kept moving.
I didn't.

I DIDN'T WANT TO GO

I didn't want to talk. I didn't want to be asked how I really was, or looked at like I was broken. I didn't want to admit how far I'd fallen. Mostly, I didn't want to be seen.

I made the appointment, but she drove. Said I'd lie if she didn't come. Said I'd say I was fine. She wasn't wrong, and it pissed me off.

There's a kind of tired that sleep doesn't touch. Where your skin feels heavy and your mind's stuck in a loop you can't even hear until it breaks you.

She talked more than I did. Told the therapist what I couldn't say out loud. I stared at the carpet, counted the fibers, and waited for it to end.

It felt like being ambushed by my own truth.

Shame mixed with relief. I didn't want to name either.

I hated her for that day. Still do, sometimes. But I know what it was—

a last-ditch move by someone who couldn't carry the silence anymore.

Call it interference. Call it control. Call it exhaustion.

That day might've saved my life.

Not out of grace.

Just because no one could ignore it anymore.

LOSS AND GAIN

California felt big—bigger than your plans, which included kids, land, and love, all built with a certainty that only youth and beer-soaked skies allow. Years later, you were counting seconds, guarding minutes, treasuring every sign that said: this is working. Wife. Children. House. Dog. The whole inventory, checked off like a life well-lived. But life doesn't implode like fireworks—it folds in on itself, quietly. The dog leaves first. Then keys. Then footsteps that used to find home without thinking.

What's left? Laughter from the backseat. A house that holds no memory. A version of happiness you're still trying to define.

The accident didn't last long, but your son's face in the mirror lingers. That reflection—part past, part new scar. And still, a new road. New truck. New laughter. Even the broken parts tell a new story.

At night, sadness has teeth. It chews through the silence. You let it. You write anyway. Not to fix it, but to name it.

Hope is fragile. But it's real. And some nights, that's enough.

A LONG ROAD TO PARADISE

They say it's a long road to paradise,
And oh, how I feel the pain.
But don't worry, they tell me,
"Everything's different now," they explain.

Yes, of course, the sun shines brighter,
And rainbows dance on the breeze,
But funny how this "new day"
Still knocks me to my knees.

The road's supposed to be smoother,
Not paved with memories and scars,
But here I am, tripping over
The same old broken stars.

"Don't you feel better?" they ask,
With their smiles so wide and fake,
As if their cheery words alone
Could fix what's left in my wake.

Oh yes, I feel so *different*,
A brand new man, can't you see?
Except for this tiny little fact
I'm still carrying me.

So I'll walk this road to paradise,
Pretending the pain's a game,
Because though things have "changed," my friend,
I still hurt just the same.

AFTER THE TREMOR

In the hush after the tremor, two currents run inside.

One trails the quake like a soft echo.
It names every crack in the glass.
It gathers shards behind closed doors.

The other leans into the wind.
It laces each breath with purpose.
It lets footsteps clear the sky.

Most of us drift in the echo,
we tend the silent rooms of memory,
we chart the tremors in our veins.

Then one day we shift—
we turn toward the open air,
we draw the storm behind us, not ahead.

In patience and small steps,
we reshape the world we carry.

TRAVELER AND THE TAMED

There was a man who traveled in time.

He wandered back and forth between past and present, future.

Time curled and bent, each round a trap which he was unable to escape.

He knew the byways, the tricks to sidestep the pitfalls,

But for some unknown reason, he still kept stumbling headlong into them.

There was a man who did not possess his mind.

They possessed power that could knock foundations and perturb the gods.

He thought that pain made him strong,

But he was wrong.

The monster found a door, opening doors to worlds beyond anything he ever imagined.

He leaped, and in the act, he found himself.

Love tempered his power, made him his own.

He was greater than a beast—he was something else.

He learned that he could heal.

The once-beast man met the time traveler.

He observed the cycles that imprisoned his friend, the same cycles that he knew all too intimately.

The once-beast man understood what it was to be hurt.

The once-beast man changed into something new, something improved.

He understood altering was not easy. It was dirty, it was hurtful, it was endless.

He knew his friend, the time traveler, would be hurt, and he promised to be there.

The beast and the time traveler went into the unknown hand in hand.

They had no maps, no warning of what lay ahead of them.

But they went on.

Because healing was the only way.

WHERE WERE WE GOING, IF NOT TOWARD EACH OTHER?

If you hadn't been there—
where would I be?
Would I have slipped through the cracks
with no one to keep an eye on me?
Would I have confused silence with comfort
and never returned from it?
There were nights
my hands trembled with nothing
and everything.
And you—
you were the only sound more deafening
than the voice that yearned for me to leave.
If I hadn't been with you,
who would've remained stationary as you unraveled?
Who would have seen past the good lie
you gave face-on?
Would you have let anyone else
cut the string and call it curing?
We weren't heroes.
We were sandbags in the other's deluges.
Dead weight, maybe.
But we held the house from drifting away.
You breathed for me
when I was ready to be finished.
I gave you silence
that didn't ask you to tell.
Maybe that wasn't love.
But it was something.
If I never came that day—
if I had said no to the coffee at night,

the long conversations,
the horrible timing—
would you still be with me?
Would I?
There's a part of you
who didn't require me.
And there's part of me
who never knew you either.
But I don't think either of them is true.
They sound like things
people who sleep well say.
You and I—
we weren't the correct story,
but we were the actual one.
And that means something.
Even now.
Especially now.
When I feel the contours of you
in the form of who I nearly became.

GENERATIONS OF FIRE

The fire crackles low,
cutting orange shapes into the dark.
The smoke wrinkles into the air
like something remembering home.

A father, his son, and his son's sons
sit shoulder to shoulder
under the slow breathing of the trees.
No one speaks.
They don't have to.

The boys stare into the fire,
eyes open with light and flinch.
The man beside them leans back,
his hand near his father's boot,
not on it, but near enough.

The grandfather watches the sparks
rise like small prayers.
He's had summers enough to know that silence can say most
often.

The wood groans
as sap gives way to heat.
The smell surrounds them —
smoke, pine, some deep old memory.
It clings to their shirts, their skin,
like memories do
when no one is straining too much.

Time is unraveled of word and clock.
Only the fire is talking
in its low, crackling voice.
It whispers something else
to each of them and somehow the same.

When the coals burn low
and the stars settle in,
they rise — together —
and return to the cabin,
with smoke, and quiet,
and something more
they can never speak.

UNSENT

I stand in the quiet,
your room without you,
beds neatly empty,
missing laughter from two.
Five years you've been mine,
my anchors, my air;
your tiny hands pulled me
from darkest despair.
I write these words slowly,
unsent but sincere;
you gave me the courage
to fight through my fear.
Your innocence shields you
from pain I hold tight—
one day we'll sit fireside
and I'll share what's right.
My demons retreat
when your smiles appear;
your laughter, your voices,
the best sounds I hear.
Time rushes past me,
you're growing too fast;
each moment without you
feels lost to the past.

I promise you both,
as true as I breathe,
you saved me, you changed me—
the best parts of me.
So keep being joyful,
keep growing so strong;
one day you'll read this
and know you belong.
Always your father,
always nearby,
anchored by your love—
my reason, my why.

TRUE FRIENDS, FOREVER

As I've grown older,
my circle's thinned.
Not by accident,
but by choice.

Friends? I have a few.
Family? Even fewer.
But the ones who stayed,
they matter.
They return the kindness,
not just take it.

They show up when it's dark,
stay when it's messy,
laugh when I can't,
listen when no one else does.

Time pulls people apart,
miles stretch too far,
and death,
well,
it does what it does.
But even then,
the real ones stay—
in spirit,
in memory,
in the echo of their voice
when I need it most.

A true friend?
That's family.
Not by blood,
but by bond.

They leave marks
that time can't fade
and distance can't erase.

They live in your heart
long after the world moves on.
Because that's what true friends do—
they never really leave.

ACROSS THE MILES, THANK YOU

I never thought I'd see the day,
That life would take you far away,
Looking back, I wish I'd known
Just how much our friendship's grown.

You've always brought out the best in me,
Seen the parts no one else could see,
Always there when times got tough,
Even miles apart, you're enough.

So thank you, my friend, for all you give,
For showing me new ways to live,
For the peace you bring when life gets loud,
And lifting me when I'm feeling down.

I miss the way you'd pull me through
With just a joke, a story or two,
And though you're far, you're here each day—
Some bonds are built to stay that way.

I wish I could call you up right now,
Just to hear your voice, somehow,
We'd laugh like it was all brand new,
And remember there's nothing we can't get through.

We've had our moments, we've had our fights,
But you're still the one who sets things right,
We're not perfect, but we don't end,
Here's to you, my true friend.

So here's to the calls and quiet nights,
To all the ways we put things right,

To you, who brings me peace from afar,
And stays close, no matter where we are.

Thank you, friend, for all the years,
For laughs, advice, and crying tears,
For being a light that never fades,
Across the miles, our friendship stays.

IN THE SHADOWS, I RISE

I stand at the edge,
The dark pulls like gravity,
Whispers of release sweeten the air,
And I wonder, for a breath,
How easy it would be
To just let go,
To stop fighting the current
That drags me under every day.

The weight of the world
Isn't on my shoulders,
It's in my chest,
Creeping in like a storm cloud,
Stealing my breath,
Turning my skin cold.
Anxiety claws at my lungs,
Depression sinks its teeth in my veins.
Together, they make me numb,
Yet burn me alive.

But then—
I hear them,
Laughing, crying,
Their tiny hands reach through the fog,
Anchor me to the earth,
Reminding me of a reason
That doesn't fade,
That doesn't shatter
Like everything else.

Some days I'm strong,
And others I just survive,
But their voices echo in the silence,
A lifeline when the dark is winning.

I pull myself up, bruised, scarred,
But not broken.

There are moments I stand tall,
Moments I sink back to my knees.
This fight doesn't end,
But neither do I.
Because in their eyes,
I see the light I can't always find within,
And that's enough.

Enough to keep breathing,
Enough to try again tomorrow.
And that's where hope lies —
In the trying, in the rising,
In knowing that every fall
Is just another step toward
Finding my strength.

IT NEVER LOOKED SO GOOD

It comes quiet at first, like a whisper,
a shadow slipping through the cracks,
a weight settling in the bones,
a voice that sounds too much like your own.

It feeds on silence, on long nights,
on the anger that curls behind your ribs,
on the moments no one sees
the ones where the mask slips, just for a breath.

Some days, you push back.
Dig heels into the earth,
claw toward something… anything… that reminds you you're
still here.

Other days, you don't.
You let it settle, let it pull,
watch the world through tired eyes
and wonder if the fight is worth it.

And then comes the moment.
The choice, the step, the fall, the flight.
One way or the other,
it never looked so good.

FOREVER IS A LONG TIME TO WAIT

The sky bleeds gold at dusk,
paints fire on the water,
but no one stays to see it fade.

I have watched a thousand sunsets,
felt the warmth spill through my hands,
only to vanish like breath on glass.

I have held names like sand,
listened to laughter turn to echoes,
left alone in rooms that once knew voices.

They call it a gift,
this long walk through endless days,
but they never ask how heavy time can be
when it will not end.

I have traced my scars like constellations,
searching for a story in them,
a reason, a purpose, a way to say:
"This is why I am still here."

But meaning is a fragile thing.
Some days it fits in my palm;
some days it slips through the cracks.

I wonder, often,
what it feels like to stop.
To step off the edge of this slow decay,
to fold myself into the quiet,
to leave before time can take me.

But the stars still burn above me,
ancient and dying,
whispering that even forever is not forever,
that even they will turn to dust.

59

So I stay.
Not because I want to.
Not because I have hope.
But because I have seen the sky bleed gold at dusk,
and it is still beautiful,
even when no one stays to see it fade.

LIVING IN BETWEEN

You don't want the silence
that dying might bring,
but each day your mind whispers,
a relentless thing.

You're trapped in this limbo,
alive but not free,
haunted by thoughts
you don't want there to be.

You carry this weight,
an ache without name,
a shadow behind you
playing a cruel game.

You don't seek escape
from life, just from pain,
yet the whispers continue,
like endless cold rain.

Days blend into nights
in a fight you can't see,
not wanting to go,
but desperate to flee.

Hope feels too heavy,
but you're holding it still,
waiting for quiet
that your thoughts never will.

You exist in this tension,
between darkness and day,
holding tightly to life,
while thoughts pull you away.

TOO MANY LETTERS

I sit to write a letter
that no one should have to pen,
to speak for a friend silenced
by battles fought within.

Words come slow and heavy,
each truth raw and clear,
trying to hold your memory,
making sure they hear.

The VA saw just numbers,
lines on paper, not the pain,
the nightmares, sleepless nights,
the shadow you became.

I knew those quiet signs,
your back pressed to the wall,
always scanning exits,
braced, expecting it all.

A cup you held too tightly,
alcohol hidden in plain sight,
your medicine and poison,
fighting darkness without light.

In writing, I'm admitting
I recognize your hurt,
because your pain mirrored mine,
same wounds beneath my shirt.

This letter is for your family,
for justice overdue,
words that hurt to share
but less than losing you.

I seal it with frustration,
hoping it reaches hearts that read,
knowing if you'd gotten help,
this letter wouldn't need.

I DOUBT I'D DO IT, EVEN IF I COULD

You asked me once,
"What would it take?"

I didn't answer.
Didn't need to.
We'd been circling the drain for months.
We both knew the sound.

You lined up the pills
on the bathroom sink.
Not a threat.
Not even a plan.
Just something to do
with your hands.

I said,
"Let's not."

You said,
"I wasn't going to."
Then paused.
"I doubt I'd do it,
even if I could."

That line stuck.
I didn't trust it.
Still don't.
But I've repeated it
some nights
like a worn prayer.
No faith in it.
Just rhythm.

Now years later,
you show up in my mirror
when I'm brushing my teeth too hard
or forgetting to eat.
Still asking,
"What would it take?"

And still,
I don't answer.
But I think it:
I doubt I'd do it,
even if I could.

THE SPACE BETWEEN THEN AND NOW

There's a silence that lingers,
not just in passageways,
but in your bones
after the flight deck cools
and the adrenaline wears off.
You talk about going home.
You picture the welcome,
the beers,
the smell of your own sheets.
But when you get there,
everything's too fast,
too loud,
and no one understands
why you flinch at nothing
and sleep too light.
Some of us didn't bring back souvenirs,
just back pain and short tempers
and a habit of checking exits
out of reflex.
We joke.
We drink.
We write things down
so we don't forget
how we became who we are

in that floating steel coffin,

in the heat,

under fluorescent lights,

just trying to hold

the pieces of ourselves together

long enough to get home

in one piece.

SUICIDE IS PAINLESS... SO THEY SAY

I sit on my bed, knife in my hand,
My life's out of control, can't understand.
A broken man, I can't let go,
Wishing to be buried, in sorrow's flow.
Haunted by things I've done and seen,
Voices whisper I should end this scene.
From night till dawn, they cause me pain,
Messing with my mind, driving me insane.
Can't sleep, cursed by nightmares grim,
Memories chase, pills lose their whim.
Life seems false, a cruel deceit,
Seeking escape, where fate and despair meet.
I sit on my bed, gun in hand,
Wanting to breathe but can't withstand.
Words engraved, a flag laid low,
Demons gleam, pulling me below.
Caught in my memories,
I can't break free,
Afraid to leave,
but it's consuming me.

STARBOARD WAKE

The ocean was angry that day.
Or maybe I was.
Hard to tell anymore.
Wind carving spray off the tops of waves
like it's trying to erase something
no one asked to remember.

Steel underfoot.
Salt in my nose.
Another day of motion without meaning.

I used to stare out here for peace.
Now I stare because it's the only place
I can look
without having to answer questions.

It's loud enough to drown thought.
But not loud enough to drown memory.
That takes work.
Takes bottles.
Takes sleep I'm not getting.

I watched the baby wave build,
then break.
Didn't even make the full roll.
Just folded into itself
like it thought better of it halfway through.
Smart wave.

The ship moves on.
Doesn't care.
Neither do the people inside.
We're all pretending this is normal.

If I jump, no one would notice
until muster.
If I don't, no one notices anyway.

Either way, the water keeps moving.
The sky stays gray.
And I go back inside,
because I'm not done yet.
Not by choice.
Just not done.

SITTING IN THE DARK

Sitting in the dark
Tears leaking from my eyes
It's been a while since I've found myself here
Not sure who came to visit and set me off
Lonely and quiet, but the quiet is crowded
Trying to drown out the pain with music
The music is helping
It's making it worse

I thought I was past this part
Done cracking open for no good reason
But here I am
Counting the spaces between breaths
Wondering if anyone notices when I disappear a little

I know I won't stay in this place forever
But that doesn't help tonight
Tonight, it hurts
And I'm tired of pretending it doesn't

I want to speak, but don't want to explain
Want to be held, but not asked what's wrong
Want to feel okay, without having to earn it

FAULT LINE

We keep a pact written in small hands,
signed on a night neither of us remembers.

You watch one horizon, I the other,
the boundary thin as breath.

Clock hands click trip wires.
Meetings, departures,
polite nods, gate open, gate shut.

Anger waits in my pocket,
a flare I refuse to strike.
Some evenings I smell the sulfur,
see a sudden glow swallow your outline,
then I let the vision cool
and count the cost in silence.

Until the ink fades,
I clip words, swallow sparks,
call restraint its own reward.

NEITHER HERE

They greeted me like nothing had gone wrong,
as if I hadn't changed, or if I had,
it wasn't something anyone should name.
I nodded, smiled, sat where I used to sit.

The walls had not forgotten how to breathe.
The questions came. I answered them in part,
enough to seem engaged but not too close.
Too much and someone might start listening.

The food was fine. The voices overlapped.
I watched the way the hours slipped around,
how even rest became another task,
another place I had to show up whole.

They meant it well. I knew that. Still, I felt
a different life pressing against my skin
the one I left, the one I hadn't shed,
the one they didn't really want to see.

This visit was a pause. That's what they said.
But nothing paused except the way I spoke.
Inside, the rhythm never really stops.
You carry noise that others call your past.

They asked me what I planned. I made a joke.
Laughter came easy, though it wasn't real.
You learn which answers help avoid concern.
You learn to exit rooms without a sound.

I wanted peace. I needed time. I tried
to be the version they remembered best.
But every version left me feeling split
too much of one and not enough of either.

They missed me, yes. I missed them, in my way.
But missing doesn't always mean return.
And leaving doesn't always feel like choice.
Sometimes, it's just the only way to breathe.

THE PULL

I stand on the edge of the old life,
its call a whisper, a promise, a lie.
Two years sober, but the ache remains,
a ghost in the shadows, murmuring my name.

It's not the world that's cruel, not today,
just a hollow ache, a path gone astray.
The weight of a life I've shaped with care,
feels like a cage when the night is bare.

I want the fire, the reckless blaze,
to lose myself in the swirling haze.
To drink, to forget, to drown the sound,
of the silent truth that presses me down.

But I know this ache, I've walked this road,
its tempting voice, its heavy load.
I'll come out fine—I always do,
but the pull tonight feels raw, feels true.

So I'll hold this moment, let it pass,
like waves that break and shatter glass.
I've built this life; it's mine to keep,
though the siren sings, I'll choose my peace.

UNSHAKEN

Not all family is given — some we find,
Woven through years, steady and kind.
Not by blood, but by something strong,
A bond that's held when days felt wrong.

You were there when the light grew thin,
A voice that pulled me back again.
Not with grand speeches, not with demands,
Just quiet strength, just steady hands.

I've seen you rise, I've seen you break,
I've seen the toll the hard years take.
Yet still you stand, still you fight,
Still you find your way to light.

So if the weight should press too deep,
If silence drowns or shadows creep,
Know this truth that will not sway —
You are not alone. Not then. Not today.

HOOFBEATS

I drive to Saratoga.
dawn hangs low, barns breathe hay and dust.
no briefing, no seats in rows,
only a round pen and one retired horse.

first step: fit panels, scrape sand flat,
feel sweat lift the fog in the mind.
second step: breathe, wait, listen
for the soft drum of hooves against dirt.

the horse tests me with one wide eye.
I keep my feet still, hands loose.
circles tighten, ears tilt,
energy trades places across open space.

rope slack; he turns to face me.
air thickens, wires hum back to life.
I rest a palm on his neck,
heat flows skin to skin, breaker clicks.

night back home tries its old tricks,
slides reels of fire and metal.
I press play on a fresh track instead:
hoofbeats, warm breath, the light stripe down his face.

this is how you change the tape.
no medals, no slogans,
just one horse, three days, a pen, a chance to choose.

if the static in your chest keeps buzzing,
make the call.
Saratoga waits, gates open,
the horse is ready to meet you halfway.

TREADMILLS

The treadmill moves
even when you don't.

The ship rolls left
your knees go light.
Then right
and gravity punches back,
twice as heavy.

Running isn't the right word.
It's reacting.
Feet adjusting,
core locking down
like the deck might pull away
mid-stride.

The machines shift,
just a little at first.
Then a swell hits
and the whole row skates
half a foot
before settling
like nothing happened.

Someone always gets tossed
legs too fast,
timing off,
belt still running
as they hit the floor.

No one laughs.
Everyone's seen it.

The air's thick,
recycled too many times.
No breeze,
no ocean smell,
just steel, sweat,
and the low whine
of a motor fighting the sea.

Outside,
heat blurs the horizon.
Inside,
you run in place,
because it beats feeling
the sway.
And sometimes
that's enough.

NIGHT SHIFT, HANGAR LIFE

Mids again.
Third week straight.
Apparently I *like* being the one who fixes shit
while the rest of the command plays house and goes home.

I don't complain, though.
Not much, anyway.
Nobody's breathing down my neck at 0300,
and the flight line's quiet
except for the APU whining
and someone trying to kick a coffee machine into submission.

My crew?
Mine, not just on paper.
Handpicked, trained by me,
and smart enough to keep busy without being told
every five damn minutes.

The chiefs?
They know better than to bother us.
They drop taskers before the shift
and magically find them done by morning,
like the hangar's got elves or something.

Even my own LPO
mostly leaves us alone.
Trusts me or gave up.
Doesn't matter which.

It's just us, the aircraft,
and the occasional bird that somehow keeps making it into the
hangar.

We knock out inspections, fix gripes,
sign off on things that were "impossible" during day shift
and still find time to argue over who's turn it is to go home
early.

I don't miss the day crew drama.
No yelling, no posturing,
no maintenance control micro-managing every wire.
Just work. Quiet. Done right.

They'll never give us credit.
Doesn't matter.
We own the night.
And the planes fly because of us.
They just don't know it.

UNDER THE WING

Steel deck sweating in the sun,
five of us pause
beneath the wing of the Hawkeye,
its engine still warm,
its belly open like it's telling us
it's tired too.

Green shirts grease-smudged,
flight suit half-zipped,
someone squints into the wind,
another smirks like
he's been awake too long
but not long enough to quit.

We don't say much.
Maybe a joke.
Maybe just the silence that fits
after the checklist is done,
after the last bolt is checked
and the danger signs still glow red.

Behind us, chains rattle,
some jet whines from the bow.
Another launch,
another cycle.

But right now—
we stand in the shadow
of the bird we keep alive,
not heroes, not models,
just sailors
doing the job
that keeps the sky full
and the mission rolling.

ALMOST HOME

Smiles for the camera.
Because that's what you do.
You square up, throw an arm around your boy,
pretend the weight isn't killing you.

Uniform's crisp.
Eyes aren't.
I see it in both our faces
the thousand-yard stare hidden behind
the "we made it" look.

We did make it.
But not without cost.
Not without ghosts.
Not without giving up parts of ourselves
we'll never get back.

This wasn't the kind of tired
that fades with sleep.
It was the kind that settles in your bones,
changes how you carry silence.

We laughed that day,
but it felt rehearsed.
Everyone wanted it to be over,
and none of us knew what to do
with the part of us that didn't want to leave.

There's a brotherhood
no civilian will ever get.
And that's fine.
But when the gangway goes down,
and we walk off separate,
no one sees how hard it is
not to feel lost.

We're alive.
We're here.
We're home, technically.
But that doesn't mean we made it all the way back.

INDIFFERENT

I'm indifferent
to a lot of things—
background noise,
opinions shouted into empty rooms,
people chasing meaning
in things I'd never look twice at.

I don't flinch.
I don't chase.
I don't correct.
Do what you want.
Say what you say.
It's your life.

Until it spills over.

Until your mess
soaks into my floor,
your noise
drowns out my quiet,
your choices
tie knots in my path.

That's when I stop being indifferent.
That's when I show up.
Not to argue—
but to draw a line.

You do you.
But not through me.
Not at my expense.

I was fine.
Until I wasn't.

CLOSED FOR SAFETY

The world went quiet,
but my head didn't.

Doors shut.
Appointments vanished.
Support systems reduced
to voicemails and buffering screens.

You don't quarantine PTSD.
You don't put depression on pause.
They don't wait politely
for the all-clear.

I watched the same walls
every night,
listened to myself unravel
in a loop I couldn't break.
No distractions.
No outlets.
Just stillness thick enough to choke on.

Some days I forgot what my voice sounded like.
Some nights I remembered too much.
Sleep was a rumor.
Peace, a joke.

People baked bread.
Learned to knit.
Reorganized their closets.
I just tried not to disappear.

And when things opened again,
the damage had already been done.

No one noticed.

They never do
until it's too late
or too loud
to ignore

36 DAYS TO LAND

Thirty-six days left.
They say we're headed to Hong Kong.
I say, who cares.
My body's done.
Back aches, eyes sting,
motivation's gone UA
and no one's turning it in.

I miss home.
Miss silence that isn't
APUs whining at 3 AM.
Miss food that didn't come
in cardboard boxes or slop trays.

I miss people
who don't talk rank,
who don't steal credit
or threaten crows
like it's some kind of power move.

Thirty-six days feels like a year.
And I know—when I get home—
everyone will say
it went by so fast.

PORTSIDE THOUGHTS

She leans on the edge
of steel and silence,
watching a world
just far enough away
to feel foreign,
just close enough
to wonder about.

The ship groans beneath her,
carrying routines,
oil,
orders,
but for this breath
none of that matters.

Behind her,
the city keeps its secrets,
stacked brick and satellite dishes,
boats tied to the shore
like they've never drifted too far
from home.

She smiles like someone
holding back a thousand thoughts,
and maybe just one:
Not long now.

THINGS SAILORS DON'T TALK ABOUT

The culture shock of coming home
relearning how to be a person again
standing in the mirror
trying to remember your face without the uniform

bruises and cuts from places that don't exist on land
scarred hands and sore backs
jeans that barely fit
still feel better than coveralls ever did

getting sick from real food
your favorite drink now fights you
getting dressed with the lights on
feels almost illegal

the weight you lost without meaning to
the silence that keeps you awake
you miss the noise
miss the motion
miss the metal groaning under pressure

you're grateful now
for clean socks
for folding laundry on a bed
for not wearing shower shoes
for water that runs hot, stays hot
for no rationing hours

you still hear it
General Quarters
Man your battle stations

you listen for the wire to catch
you wait for it to retract
you breathe when the pilot is safe

Man Overboard
and they didn't say it was a drill
everyone moves quicker this time

Evening prayer
Fire, Fire, Fire
they're always catching fire
someone always hits their head
you stopped noticing when it stopped hurting

Medical emergency in compartment…
who's in that compartment?
do I know them?
I hope it's nothing

you forgot what it's like to be alone in a room
for more than ten minutes

you come home in a different season
the corn is gone
the grass is green
the baby was born
bless her heart

where was I?
it's been years
was I even gone?

First call.
First call to colors.

Coming home from all of this
will be so hard
so easy

FINAL WATCH

I've cut through ocean's restless tide,
where steel and salt embrace the sky,
And soared beyond where eagles glide,
Into the boundless, blackened high.
Through whispers of the midnight wind,
I've sailed on wings both tried and true,
In silent arcs where stars rescind,
Their light to those who fade from view.
Beyond the thunder's ruthless roar,
I've fought through storms with iron will,
Survived the trials, weathered more,
And rose from depths no calm could still.
Yet there, above the twilight's breath,
Where waves and clouds forever part,
I faced the dark, the void of death,
And felt its coldness pierce my heart.
Now drifting where the heavens weep,
No compass, chart, or beacon's guide,
I find in shadows' quiet keep,
The solace sought by those who've tried.
This final flight, into the deep,
Where loss and hope together blend,
I leave behind the world's embrace,
To join the stars, my journey's end.

WORN OUT FACES

Something inside me died.
(Not all at once. Just... pieces.)

Please don't go.
I want you to stay.
But I never said it out loud, did I?
Only in the silence
after the door shut.

You were there
and I forget the rest.
It's all fog
faces with no names,
rooms with no time.

You can't rewrite the past.
But damn,
don't think I didn't try.

You were always so lost in the dark.
I saw it.
I just...
thought I could be light.
Thought maybe standing next to me
would be enough.

Open your eyes.

No, not like that.
Not the hospital kind.
Not the pale ceiling kind.
I mean, open them and *see* me.

Worn out faces
in the mirror,
in the photo,
in the memory I wish would let go of me.

I still talk to you.
You don't talk back.
I still wait up.
You don't come home.

Maybe something inside you died too.
Long before you left.
Long before I noticed.
Maybe I noticed
and turned away.

INTO THE EMPTY BLUE

Three empty chairs
the flight line whispers quiet now,
tools in hand hesitate,
turn slower, doubt shadows every glance.

You know the odds
but knowing isn't preparation;
the names announced, the casual voice
breaking something inside, forever.

Memories surface
port drinks, salty jokes,
brief words through a duty window,
a promise left unfulfilled.

The silent question lingers
in metal inspected, cables checked
hands trembling, wondering
if attention slipped
when attention mattered most.

Everyone says,
"You couldn't have stopped it,"
but doubt is louder, clearer,
always replaying what-ifs.

And still,
planes launch again,
fear pushed aside
by necessity, by trust
frayed but intact,
in each other,
in routines.

Some wounds
never close fully,

just ache less often,
reminding you
of friendships, laughter,
sunglasses indoors
on mornings too bright
for anyone to bear.

What remains are memories
goofy grins, country songs,
and the quiet salute,
Fair Winds and Following Seas,
given softly,
into the empty blue.

THRESHOLD OF LAND AND SEA

Sometimes you find yourself
standing on the threshold of land and sea,
wondering how you got there
and what you're supposed to do now.

It's not the first time you've been here,
though the shape of the shoreline has changed.
The tide always moves differently,
the waves never break in the same rhythm twice.
You've noticed that much.

Behind you, the land is solid,
its weight pressing against your back—
the memory of footsteps,
the miles you've carried,
the people who spoke in voices
that still echo in your mind.
Some were kind.
Some weren't.

Ahead, the sea.
Wide and restless.
It doesn't care for what you've left behind.
It doesn't care about you at all.
And maybe that's the hardest part.
Or maybe it's the easiest.

You dig your feet into the wet sand,
as if looking for answers in the sediment.
But the sand gives way,
like everything else.
And the question still hangs there:
What now?

You could turn back.
You've done it before.
The land is familiar, if unkind,
and there's comfort in the known shape of suffering.

You could step forward,
let the sea swallow you whole.
Not in surrender,
but in the hope
that maybe the water holds something
you've been missing.

Or you could stand here awhile,
letting the wind push at your shoulders,
letting the spray sting your face,
letting yourself feel what it's like
to be caught between two worlds
without the need to choose—
not yet.

Because sometimes the hardest thing
is not moving.
Sometimes the answer comes in the waiting,
in the pause between tides,
in the sound of waves
breaking against who you were,
reshaping who you might become.

And so you stand,
wondering how you got here
and what you're supposed to do now.
Not realizing, perhaps,
that standing is already enough.

WHAT HOLDS

There are days I count my value in negative space.

If I were gone, would the air feel thinner? Would the sky blink differently? Probably not. The mail would still arrive, traffic would still jam up, someone would still complain about the weather. The planet would spin the same indifferent spin, unbothered by my absence.

I know what this voice is.

I've studied it. Read the books, heard the therapists. "That's depression talking," they say.

And they're right. But knowing doesn't make it quieter.

It just means I can name the thing clawing at my throat.

People say, *"But your kids."*

And they're right too.

My kids are the gravity keeping me here, the only proof I have that I still matter in real time.

They don't see the storm behind the stillness, not always.

But they laugh, and something in me flinches—something buried and human, that wants to be worthy of their eyes.

They need me now.

But I know the math.

One day, they won't.

What then?

What do you do when the only tether is scheduled to loosen?

Sometimes I rehearse the argument: "People would care."

Yeah, they would. For a bit.

Then the world would eat their attention again. Bills. Laundry.

Deadlines. Birthdays.

The soft tragedy of someone disappearing doesn't last long in a culture built for speed.

But I'm still here.

Not because it gets better. That's a story people love to tell.

I'm here because I'm stubborn.

Because I'm curious.

Because sometimes staying is the only way to spite the thought that says I shouldn't.

Some nights I lie in bed and pretend I'm a tree. Not one people carve hearts into or hang swings from. Just a tree.

Standing. Useless. Present anyway.

Maybe that's enough.

I want this to mean something.

I want someone else—maybe you, reading this—to feel the quiet echo and recognize it.

Not to fix it. Just to know it exists.

That someone else woke up and wondered why.

That someone else didn't answer the phone because they didn't want to lie.

That someone else made it through a day they didn't think they would.

We don't talk about this enough.

So this is me talking.

I'm still here.

And if you are too, then maybe that's the whole point.

PICTURES OF YOU

I've been looking so long at these pictures of you
that I almost believe that they're real,
captured fragments of moments held like fragile glass,
scenes that rewind in the dark like fading reels.

Each image is vivid, stubborn and bright,
as if time had paused to cast you in light.
But the colors bleed, slipping between,
moments I held, and those I dreamed.

Each echo I hear is softer than the last,
as the outlines blur in a landscape of past.
Every glance, every laugh—now hollow and thin,
like wisps of smoke that the night pulls in.

I reach out, grasping at air and regret,
tracing the edges of things I forget.
But you're there, just beyond the veil of sleep,
where shadows keep secrets and memories keep.

How long can I wander these halls of might-have-been,
before the echoes fade to whispers unseen?
I've been looking so long at these pictures of you—
at ghosts that won't leave, yet will never be true.

HURT

I hope whatever tore you apart and made you tear my heart from my chest finally leaves your mind.

I hope the pain that broke you and bled into me teaches you to love deeper.
If my shattered heart helps you heal, I hope it's worth it. I hope it brings you peace, a quiet moment amidst the storm, a chance to breathe again.

I hope it unlocks the dungeon where your monsters hide— the ones you've refused to face. I hope you find the courage to look them in the eyes, to understand them, to conquer the fear that has held you captive.

Maybe my agony will help you defeat them. Maybe my tears will wash away the darkness that clings to your soul.

When you walked away, you set a fire in my chest, burned every trace of love we had. The flames consumed my hopes, turned our memories to smoke, left nothing but ashes behind. But maybe, in those ashes, a flower can still bloom— proof that love existed, even if it wasn't enough. Maybe it will be a reminder that love, once planted, can grow again, even in the most desolate places.

I hope that one day, you can look back without pain, that you can see the beauty In what we shared, even if it was fleeting. I hope the scars we gave each other become stories of survival, of growth, of learning to love despite the hurt.

And I hope, more than anything, that you find a way to love yourself. Because that's where the healing starts, and that's where new love can begin

I REMEMBER YOU, PART 2

(*Older self — slow, calm*)
I remember you.

> (*Younger self — dry laugh*)
> That makes one of us.

(*Older*)
Sitting in the dark, back against the wall.
Numb.
You said it was peace, but it wasn't.
It was the absence of anything else.

> (*Younger*)
> It was quiet.
> That was enough.

(*Older*)
No, it wasn't.
You were suffocating in the quiet.
Drinking to silence the noise you wouldn't face.

> (*Younger, biting*)
> You think I wanted this?
> You think I asked for any of it?

(*Older*)
No. I think you broke under weight no one saw.
And I think you thought that was weakness.

> (*Younger*)
> It was weakness.
> Every day was a failure I couldn't explain.

(*Older*)
You called it strength—keeping it all in.
But it was fear.
You couldn't even look in a mirror without seeing ghosts.

                      (*Younger* — *quiet*)
                      I still see them.

(*Older*)
So do I.
But I've stopped letting them speak for me.

                      (*Younger*)
                      They never shut up.

(*Older*)
They don't have to.
You just have to stop answering.

                      (*Younger* — *sarcastic*)
                      That simple, huh?
                You act like you're so far above me now.

(*Older*)
Not above.
Past.
I walked through everything you left behind.
The shame. The guilt. The nights you almost didn't survive.

                      (*Younger*)
                I didn't think we would survive.
              Every morning I woke up surprised.

(*Older*)
And every morning, you still got up.
Even when you hated yourself for it.

                      (*Younger*)
                I hated myself for everything.
              For surviving when others didn't.
     For breaking when I was supposed to be the strong one.

(*Older*)
You were strong.
Not in the way you thought, but strong.
You lived long enough to let me exist.

> (*Younger*)
> I didn't want you.
> I wanted silence.
> I wanted out.

(*Older*)
I know.
And I hated you for that.
For a long time.

(*Pause.*)

(*Older*)
But now I see you.
The broken, bleeding, angry version of me.
And I don't want to destroy you anymore.

> (*Younger*)
> Then what do you want?

(*Older*)
To let you rest.
You did your part. You carried the pain.
You kept the body alive, even when the mind fractured.

> (*Younger — voice cracking*)
> I didn't know how to ask for help.

(*Older — soft*)
I know.
And I didn't either.
But I learned.

(*Younger*)
Do you still hear it?
The hum in the back of your skull? The heaviness in your
chest?

(*Older*)
Some days.
But I know what it is now.
And I don't run from it.

(*Younger — quiet, almost a whisper*)
I was scared.

(*Older*)
I still am.
But I'm not hiding.

(*Younger*)
You don't hate me?

(*Older — firm*)
I hate what you went through.
But no—
I don't hate you.
I'm here because of you.
But you don't get the wheel anymore.

(*Younger — long pause*)
Then let me sleep.

(*Older*)
Rest.
I've got it from here.

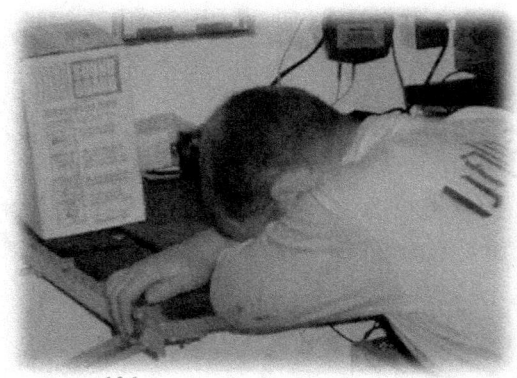

BEFORE YOU GO

These poems carry pieces of my memory, but their meaning isn't fixed. They're stitched with silence, doubt, hope, and the kind of truths you only whisper to yourself at 2 a.m.

What they stirred in you matters as much as what they meant to me. If you're willing, here are a few questions to sit with — when the last page fades, but something still lingers:

- What version of yourself did you recognize in these pages — and what part of you are you ready to leave behind?

- Which silences spoke the loudest — and what truth might be hiding inside them?

- If you were to write a poem called *I Remember You*, who would it be about — and what would you need to say?

Want to Share Something?
If something in these pages stirred something in you — and you feel like sharing it — I invite you to write me at: iremeberyoubook@gmail.com

I can't promise a reply, but I will read every message.
If nothing else, know this: your story matters. Your silence mattered. And if you're still here, *that* matters most of all.

MENTAL HEALTH RESOURCES

For anyone struggling with PTSD or mental health challenges, there are resources available to help. You don't have to face these battles alone. This is the most current information available at the time of publishing.

If you or someone you know is struggling or in crisis, you don't have to go through it alone. These resources are available 24/7:
- **988 Suicide & Crisis Lifeline**
  Call or text 988 for free, confidential support.
- **Veterans Crisis Line**
  Call 988, then press 1
  Text 838255
  Or chat online at <u>VeteransCrisisLine.net</u>
- **Crisis Text Line**
  Text HOME to 741741 to connect with a trained crisis counselor.
- **National Alliance on Mental Illness (NAMI)**
  Call 1-800-950-NAMI (6264)
  Or visit nami.org/help for support and education resources.
- **SAMHSA's National Helpline**
  Call 1-800-662-HELP (4357)
  Free, confidential help for mental and substance use issues, available in English and Spanish.
  These services are anonymous, free, and available anytime.

Reaching out for help is the first step toward healing. You don't have to go through this alone.

ABOUT THE AUTHOR

(P)

Patrick J. Hughes is a poet, Navy veteran, and still-figuring-it-out imperfect human. He has written fiction before, but never a collection of poetry. His writing usually starts as something that he must exorcise from his brain. Some of it turns into stories. Some of it turns into poems.

He served more than a decade in the United States Navy, weathering deployments, duty days, and the off-radar wars no one acknowledges. After serving in the military, he continued in the defense community through employment in the defense industry before redirecting himself to healing, growth, and writing.

A disabled vet these days, a full-time college student, and the father of twin sons who made life messier, noisier, and more richer than he ever could have envisioned, Patrick believes words are important—not to fix everything, but to declare truths that are unuttered.

For more of his writing, military experience, and mental health journey, visit his blog at https://funkyotter.blog.

www.ingramcontent.com/pod-product-compliance
Lightning Source LLC
Chambersburg PA
CBHW070343130626
46556CB00007B/3000